SAFETY IN OUR PROFESSION

¿HOW CAN WORKING SAFELY HELP YOU RAISE YOUR STANDARDS AS A PROFESSIONAL?

7 PRINCIPLES TO BE A PROFESSIONAL IN AVIATION WITH A DISTINCTIVE TOUCH.

Written by: Edwin Fernando Barreto Salamanca

Property of Edwin Barreto. - All rights reserved. Without limiting the reserved copyright rights, no part of this book may be reproduced in any form or by any electronic or mechanical means, including information storage and retrieval systems, without the written permission of the author. The only exception is the mention of a commentator, who may quote brief passages in a review.

WWW.AVIACIONAEROED.COM

WARNING

This book is designed to provide information and motivation for its readers. It is sold with the understanding that the author is not engaged in providing any kind of psychological, legal, or any other professional advice. The instructions and advice in this book are not intended to be a substitute for guidance. The content of each chapter is solely the expression and opinion of the author. There is no express or implied warranty by the choice of the editor or the author included in any of the contents in this volume. Neither the editor nor the individual author shall be liable for physical, psychological, emotional, financial, or business damages, including, without limitation, special, incidental, consequential, or other damages.

INTRODUCTION

Welcome to this guide to strengthen your skills as a professional! Here you will find various techniques that will allow you to stand out among others in the field of aviation. Remember that there is no single absolute truth, but I am sure this information will be useful to you if you decide to apply it.

The main focus of this document is to create a safety culture in the aeronautical environment, starting from each individual, that is, from our BEING. Keep in mind that what you will read here is not intended to be an absolute truth but rather some principles that, in my personal and professional experience, have modified and improved my life.

Don't worry about following every word verbatim; the important thing is to adapt all the information you find here to your reality. You decide which aspects to take and which ones to discard completely. You may not always agree with what I say, and that is entirely valid, as what I present here is my perspective on how to generate a safety culture in individuals and organizations in the field of aviation.

I recommend not rushing to read the entire book at once. It's best to review each chapter and reflect on how it can be applied to your profession.

My goal is to provide you with information that has been useful in my professional life and that I believe will also be helpful to you if you choose to. I ask that you keep an open mind to learning and enjoy the process. If you have invested in this book, it's because something I've said has resonated with you, and I know that part of its content could support you in making a significant change in your life.

Remember that success is a process, not a destination.

I hope to accompany you in your professional and personal growth, being the support you may need in your life. Allow me to share part of my life purpose with you:

To be an inspiration for the person I become every day, positively impacting lives.
Edwin Barreto

I know I still have a lot of work ahead to fulfill my life purpose, but the important thing is that I am already on the right path since I started writing and compiling with all my heart the information presented in this book for you.

I work with dedication and will continue to do so because I love helping people who seek my support with the knowledge I possess, whether it's little or much. I hope you enjoy the reading you are about to start.

Success in your personal and professional growth!

CHAPTER 1
THE 7 PRINCIPLES IN OUR PROFESSION

We will begin by clarifying why I have determined that there are seven principles that we should incorporate or maintain in our professions. Let's carefully examine each of them and how they can be seamlessly integrated into our lives, understanding how these principles crucially support our future. Upon concluding the book, you will find two key principles that form the foundation that has assisted me in building and sustaining these principles in my life.

YOU HAVE THE POWER TO BE THE BEST TODAY AND ALWAYS.

It is important to consider that these principles may seem both easy and challenging to fulfill and incorporate into life, but it solely depends on the information you are assimilating in your mind at this moment.

The real questions you should ask yourself are:

- Have I already incorporated the 7 principles into my life?
- Do I need to incorporate or improve these 7 principles in my life?
- Will I be a differentiating professional if I uphold the 7 principles?

The answers to these questions are solely within you.

Throughout this book, we will review each principle and how I have incorporated them into my life and organizations to generate a culture of safety. I invite you to evaluate each of these principles one by one and consider if you need to make any adjustments in your profession or organization.

CERTAINLY, PLEASE GO AHEAD AND MENTION THE SEVEN PRINCIPLES. I'M HERE TO ASSIST YOU FURTHER.

1. The initiative
2. Being a problem solver
3. My honesty
4. Teamwork
5. Vision with continuous learning
6. Individual responsibility
7. Effective communication

Throughout the book, we will delve into each of these principles and how to apply them in our professional life to stand out and make a positive impact on our career. Remember that the key lies in the practical implementation of these principles and how they become an essential part of who we are and how we operate as aviation professionals.

CHAPTER 2
PRINCIPLE 1: INITIATIVE

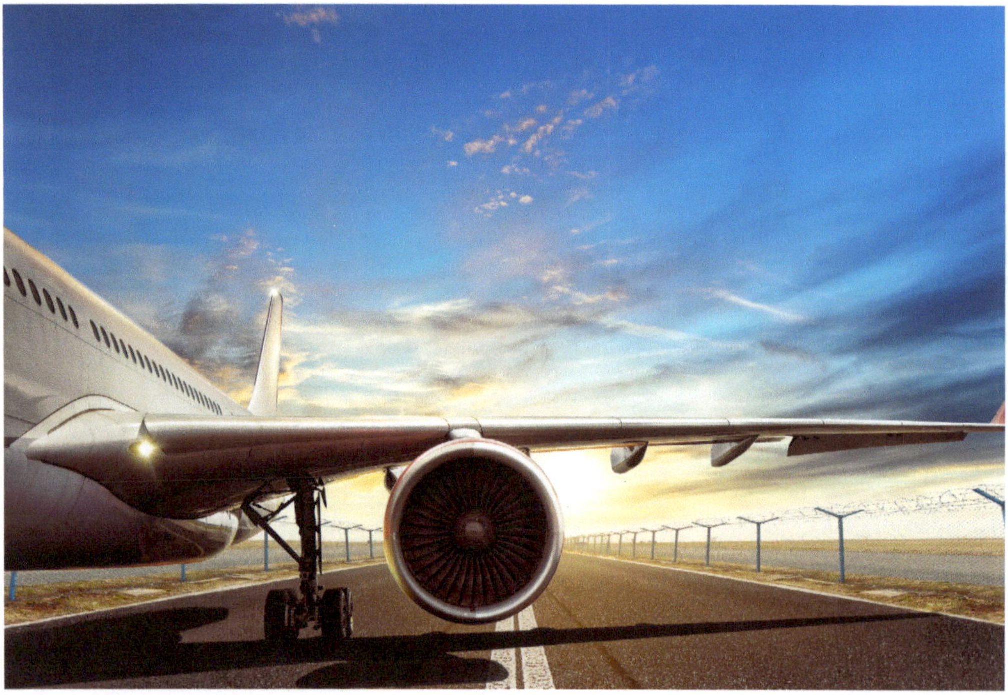

THE PARABLE OF THE ORANGES

Allow me to introduce you to this powerful principle through a parable that made me reflect on my role in a company. This story changed the way I work, and I hope you share it with your colleagues and teams if you are leading them.

I call it **Missing Oranges,** alluding to the need to go a little further in any task we undertake.

Let's read the PARABLE for a better understanding.

There was a young man who had ambitions to work for a company because it paid well and had a great reputation. He prepared his resume and had several interviews. Finally, he was given an entry-level position. He then turned his ambition into his next goal: a supervisor position that would grant him more prestige and a higher salary. So, he completed the tasks assigned to him. He would come in early some mornings and stay late so that the boss would see him putting in long hours.

After five years, a supervisor position became available. But, to the young man's great dismay, another employee who had only been with the company for six months got the promotion. The young man was very angry and approached his boss, demanding an explanation.

-The wise boss said:

Before answering your questions, would you do me a favor?

- Yes, of course, said the employee.

- Would you go to the store and buy some oranges? My wife needs them.

The young man agreed and went to the store.

When he returned, the boss asked:

What kind of oranges did you buy?

- I don't know, the young man replied.

You just said to buy oranges, and these are oranges. Here they are.

- How much did they cost? the boss asked.

- Well, I'm not sure, was the response.

You gave me $30. Here's your receipt, and here's your change.

- Thank you, said the boss.

Now, please, take a seat and pay close attention.

Then, the boss called the employee who had received the promotion and asked him to do the same task.

He readily agreed and went to the store.

- When he returned, the boss asked:

What kind of oranges did you buy?

- Well, he replied, the store had many varieties: there were navel oranges, Valencia oranges, blood oranges, mandarins, and many others, and I didn't know which type to buy.

But then I remembered you saying that your wife needed the oranges, so I called her. She said she was having a party and was going to make orange juice. So, I asked the grocer which of all these oranges would make the best orange juice. He said Valencia oranges were full of very sweet juice, so that's what I bought. I left them at your house on my way back to the office. Your wife was very pleased.

- How much did they cost? the boss asked.

- Well, that was another issue. I didn't know how many to buy, so once again, I called your wife and asked how many guests she was expecting. She said 20. I asked the grocer how many oranges it would take to make juice for 20 people, and it was a lot. So, I asked the grocer if he could give me a quantity discount, and he did! These oranges usually cost 75 cents each, but I only paid 50 cents.

Here is your change and the receipt.

- The boss smiled and said:

Thank you; you may go.

He looked at the young man who had been observing. The young man stood up, shrugged, and said:

I see what you mean, as he left the office disheartened.

What was the difference between the two young men?

Both young men were given the same task, but one of them went above and beyond and paid attention to details. The difference between the two employees was that the first one was motivated by money, position, and prestige. The second young man was driven by an intense desire to please his employer and an internal commitment to being the best worker possible, and the result was evident. This parable teaches us that anyone can become a great employee if they desire it, and the intention for their growth should be driven by their aspirations to become a better person. Great employees should always be willing to help their co-workers become great leaders.

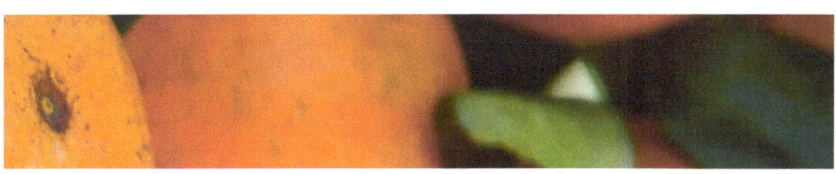

GO THAT EXTRA MILE IN EVERYTHING YOU DO!

With this powerful parable, I discovered how to push myself to give more without being asked, a fundamental principle for everyone in organizations, especially for us professionals who, at times, expect more without giving more.

Initiative involves offering that extra mile that many are not willing to give because they believe everything should always come at a price. However, I understood that if I wanted to stand out as a professional above the average, I had to deliver more than what was asked of me. Thus, I experienced a myriad of positive results in my learning and experience.

By taking two or three steps ahead in every task and job we undertake, we prepare ourselves to easily handle future situations and gain a deeper understanding of our activities.

Since I incorporated this parable into my life, I have always maintained the initiative to go that extra mile, not because I expect something in return, but because I understood that it provides me with invaluable learning. I cannot deny that doing so has resulted in greater financial gains and significant growth in my professional life (although it doesn't always guarantee this, it certainly always leads to more learning).

I always visualize myself as the owner of the best aerospace company in the world, imagining the type of people I would like to work with. People who are team players and who will take our organization to the highest level, always working side by side. I always come to the same conclusion:

I want to work with the best!

Imagine for a moment that you have a company, a great empire; surely, like me, you would like to have the best professionals by your side. I am sure that all entrepreneurs want to have the best people in their companies. So, why do we sometimes not strive to be the best at what we do?

It's not about a lack of knowledge, but sometimes, we take actions that are not the best because we have grown tired of what we do, feel that we are not valued enough, or are trapped in a comfort zone that prevents us from giving our best. However, we cannot fall into complacency, as I am sure that you, as the owner of a company, would not want employees who do not strive to give their best.

Perhaps some may argue that it's a lack of motivation on the part of the company, but that's an excuse from a mediocre professional. Remember that our profits are directly related to the value we bring to the market. If we are not currently generating the desired income, is it really the fault of the organization we work for? Or should we strive to become more valuable to the company?

Motivation doesn't come from outside; we must find it within ourselves. It's true that in some organizations, they may not appreciate us as they should, but is that a problem with the organization, or is it our responsibility to stay in the same position for years without progressing and only complaining?

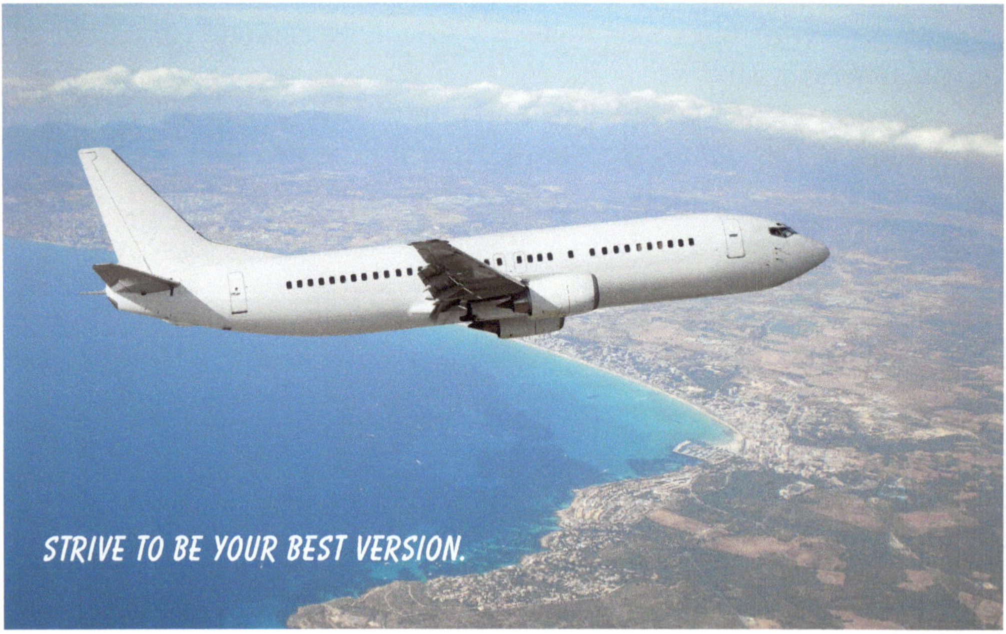

STRIVE TO BE YOUR BEST VERSION.

It's time to evaluate what we do every day and realize that what we often lack is having this first key: INITIATIVE. It's about becoming someone better, having something better, learning continuously, becoming the differentiating professional, the **'BEST SELF.'**

I know that some of the ideas I present in this book may cause discomfort or hurt feelings, but let's be realistic: do you think the world will move forward because you or I are not doing well?

The answer is NO!

No one will do anything for us; we are responsible for taking the initiative to move forward, to surpass ourselves, and to obtain what we desire. Let's set aside the disease of excuses; they lead us nowhere. Let's stop justifying ourselves by saying that we are not good because we are not paid what we deserve. Let's analyze what we are truly doing.

I conclude with this elaboration on this valuable key: *always give your best, regardless of the salary, what others say, or what happens. Be assured that life will return the same as you offer, perhaps not immediately, but life leaves nothing unrewarded. This principle has allowed me to keep my good name high.*

I am convinced that this key can help you, just as it has helped me. Becoming the best in our profession requires dedication and giving more than is asked of us. Let's commit to always going that extra mile.

GO THAT EXTRA MILE TODAY AND EVERY DAY.

Remember: **NO ONE WILL DO THINGS FOR YOU OR FOR ME.** We will be responsible for incorporating initiative into our actions.

Now that you have concluded the reading of this first key, you can take a moment to reflect on your actions and make adjustments if you deem them necessary. Or you also have the option to close this book and never open it again, but I invite you to think about those around you; perhaps they also need to know this information, and you can support them in understanding the principle of **INITIATIVE.** Sometimes, we assume that everyone should know it, but the reality is different.

Message to the Company

The safety culture in organizations is strengthened by incorporating the principle of 'INITIATIVE' from top management. Let's acknowledge that many employees need a push from CEOs to reach their full potential. Pretending that employees will always be the best just because it should be so is a naivety that can be found in the upper echelons of the organization. Remember that humans need recognition to stay motivated and keep moving forward.

Companies can embrace **INITIATIVE** to enhance the working conditions of their staff. I don't just mean economic aspects but maintaining a healthy culture regarding employee treatment. When we provide the respect people deserve, there's no need to remind them to give their best; it arises naturally.

CHAPTER 3
PRINCIPLE 2: BEING A PROBLEM SOLVER

Always, we will face situations to be resolved in our lives, and the way we confront them sets us apart.

Let's remember this:

They are not problems, they are situations to be resolved.

A technique that has always helped me solve situations is trying to see a broader picture of what is happening. In life, everything happens for a reason, and this is where we must evaluate why certain things are happening to us.

READING ACCELERATES TIME.

My approach has always been to present at least 5 possible solutions to any situation that arises. Although sometimes I may only manage to think of 3 or even 1, this will depend on the circumstances. (Ideally 5 or at least 3)

To be a great problem solver, I incorporated two key strategies into my life. The first is the habit of reading, but not just any reading, focusing on material that helps me grow as a professional, like this book you are reading now. Knowing the stories of other people and the possible consequences of their actions has allowed me to avoid repeating mistakes.

The second strategy is to establish relationships with people who are experts in different subjects. Although I cannot know everything, I have learned to identify individuals with skills complementary to mine and create a network of mutual support.

I acknowledge that I previously had a cold attitude towards people in need, but that changed when I realized I needed to be more human and build genuine relationships. Now, I take pride in being able to support others without expecting anything in return. I understand that this impacts our relationships, and I always seek to **give my best from the heart.**

To be a problem solver, it is crucial to face different situations and learn from experience. Likewise, maintaining an open mind to learning makes it easier to confront life's challenges.

By becoming a great problem solver, I have become attractive to organizations because no company desires professionals who only present problems. Everything changes when we focus on being solvers; our lives become simpler, and problematic situations are reduced.

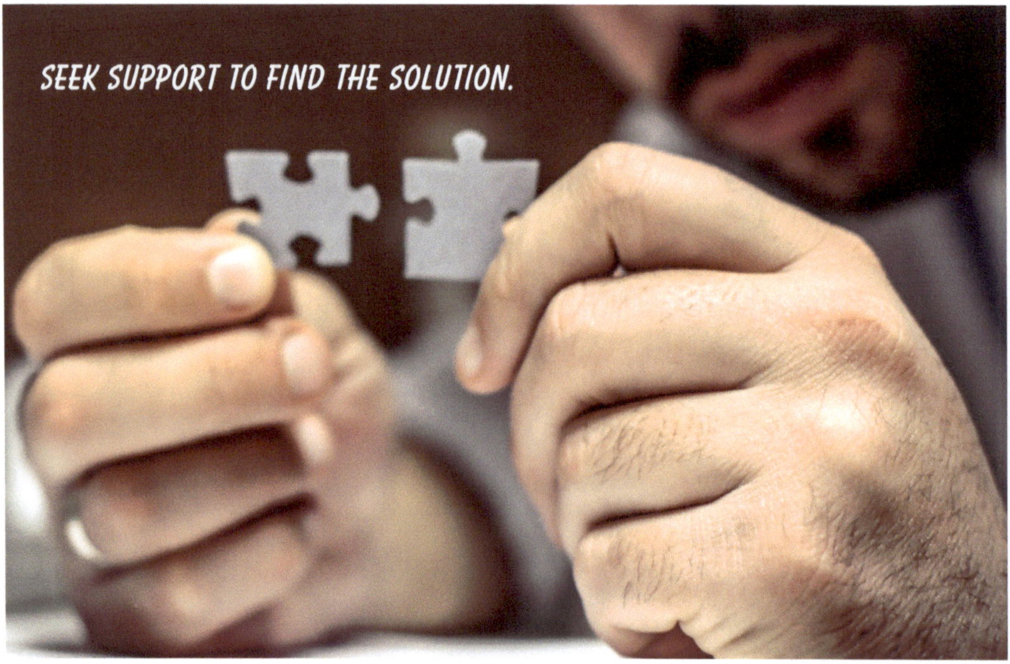

SEEK SUPPORT TO FIND THE SOLUTION.

Surrounding ourselves with people with complementary skills and being supporters for others enriches us both personally and professionally.

This book seeks to contribute a grain of sand to your profession based on my experience. But how do you make it yours? Start by identifying pending situations to be resolved and analyze how you can address them. Although it is not easy, if you commit to being a great problem solver, your life will improve.

Remember that no one will solve our situations for us; we are responsible for our lives.

Message to the Company

Creating a safety culture in an organization involves instilling confidence in workers to become excellent problem solvers. As leaders, we cannot hand everything to them passively; we must provide them with the necessary tools to grow and find their own solutions.

When a worker presents a situation to be resolved, it is an opportunity to involve them in the process. Ask that person what they believe would be the best solution from their position. You will be surprised by the amount of information and excellent ideas they could contribute.

The workers on the front line know in detail what happens in their positions. If you have employees who do not propose solutions, it is your responsibility as a leader to guide and motivate them to be able to offer ideas. Help them develop their problem-solving skills.

Personally, when situations arose to be resolved, I used to ask my team what could be the possible solutions and allowed them to arrive at the best option themselves. Even if I already had a solution in mind, I encouraged them to learn to find answers on their own.

Being a great leader involves teaching people to do what they need to do and not being indispensable in the area. Our team must be able to solve problems even in our absence. If not, we have a situation to address as leaders.

HAVE A SPARE TIRE INFLATED.

A phrase that has supported me in this approach is: **'Give a man a fish, and he will eat for a day; teach him to fish, and he will eat every day.'** We must empower our team to be self-sufficient in problem-solving.

If we cannot step away from our office without being called, that indicates that the safety of the company is at risk. If a leader is absent, the company should not be affected because the team must have the ability to continue their work.

I learned from a great leader the importance of having our **'spare tire inflated.'** This means having people on our team capable of handling any situation that arises, so the company is not affected if we are not there.

If we ever lose our job because someone we taught has surpassed us, instead of resenting it, we should self-assess and strengthen our weaknesses to grow as professionals. We must learn from each situation and see both sides of the coin to adapt the best to our lives.

Remember that everything in life happens for a reason. Review what benefits you the most from this information and use it for your professional growth.

Saving money with this principle:

As a company: Teach your leaders to promote problem-solving skills in their teams. This will save time, effort, and resources for the organization.

As an individual: Examine your situations and study how others have solved similar problems. Learn from their experiences and move forward. I assure you that you will save time and money by making informed decisions.

ARE YOU GROWING WITH EVERY SITUATION IN YOUR LIFE?

> *When you know what you want, and you want it bad enough, you'll find a way to get it.* – Jim Rohn

WWW.AVIACIONAEROED.COM

CHAPTER 4
PRINCIPLE 3: MY HONESTY

The value of being an honest and reliable professional

A true professional is distinguished by the quality of their character and the commitment they undertake, but there is a crucial element to maintain this over time:

**NEVER COMPROMISE YOUR GOOD NAME,
BECAUSE RECOVERING IT IS VERY DIFFICULT.**

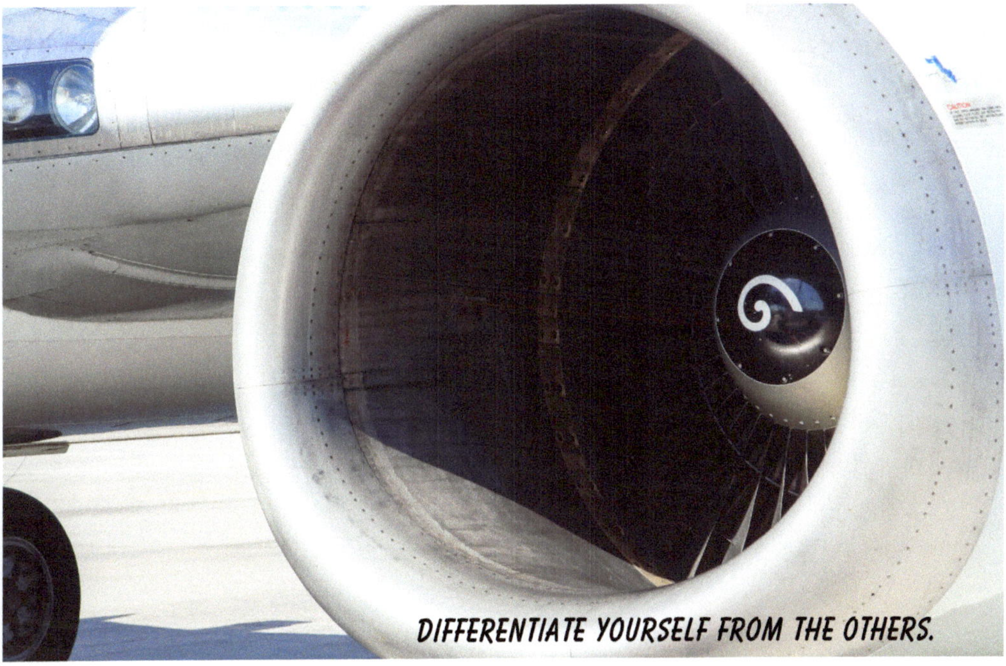

DIFFERENTIATE YOURSELF FROM THE OTHERS.

To ensure that we are respectable professionals, we must ensure that we fulfill our commitments to everyone. If we commit to something, we must fulfill it, as having the courage to keep our promises sets us apart.

If we are not willing to fulfill, it is better not to tarnish our name and simply refrain from committing. Every time we fail to fulfill what was promised, our honesty is affected, and that should not happen.

Honesty is a value that I have always worked on to be a reliable person in any organization, and there is no need to lie in any circumstance.

Throughout life, we may face situations that test our honesty, and although we may not want to damage it, it is vital to be prepared to face any challenge that arises.

The seven principles I present are interdependent, and by incorporating them into our lives, we stand out as exceptional professionals.

We can maintain our reputation intact by learning from people who have faced situations similar to ours and by developing skills to handle any situation. **Constant preparation is key.**

I recall a relevant quote from Warren Buffett:

**Honesty is a very expensive gift,
don't expect it from cheap people.**

GAINING TRUST IS THE BEST THING IN THE WORLD.

I have never considered myself a person without value; on the contrary, I see myself as an exceptional person, and I'm sure you are too. Therefore, it is our responsibility to gift everyone with our HONESTY, **which generates trust.**

I won't lie to you; there were moments when my honesty wavered, and recovering my good name with those I disappointed has not been easy, but I am gradually achieving it. **It's not easy!**

It's true, not everyone is willing to hear our honesty, and this can lead to uncomfortable situations and dissatisfaction, but each person must decide how to act because, just as we expect honesty from others, we must also be honest.

This process is not easy, I know from experience, but I always find a way for honesty to prevail, and since I incorporated it as a principle in my life, I have gained the trust of many people.

As professionals, we must trust people and their honesty when they speak. Surely, you have experienced disappointment when someone breaks their word, haven't you? So why make others feel the same? Our honesty has not always been impeccable, but I have worked every day to make it a part of me, and I hope you do too.

Our profession in aviation demands the utmost honesty because the lives of many people are at stake. Any lie, no matter how small, could contribute to an aviation accident in the future. The trust we generate by being honest is vital.

> **GAINING TRUST IS THE BEST THING IN THE WORLD.**

Facing difficult situations with decisiveness and honesty is preferable to regretting later for avoiding the truth. If something does not make you happy in your current life, be honest with yourself and make the necessary decisions.

'NO ONE WILL DO IT FOR YOU.'

In aviation, honesty is the most valuable gift we offer to all those who fly daily. Trust in the aviation industry is based on the honesty of everyone involved.

Passengers trust airline companies due to the honesty reflected in the maintenance and defenses of the company to ensure safe operations.

Pilots and crews trust in the proper maintenance of aircraft.

Companies trust every professional within the organization to maintain safe operations and comply with safety standards.

Remember! Honesty is the foundation upon which we build strong relationships and become trustworthy and valuable professionals.

WE ARE RESPONSIBLE FOR OUR ACTIONS

Message to the Company

Promoting an environment of trust and honesty among workers is a shared responsibility between top management and middle levels of the organization.

If as a leader you feel uncomfortable with the honesty of your employees, it is crucial to reflect on the true reason for that feeling. On certain occasions, I found myself in that position, and although at first I disliked receiving honest feedback, I learned to restrain myself and listen carefully. I would take deep breaths, even have a coffee to relax, and then analyze my reaction.

When I approached the situations with calmness and objectivity, I always reached the same conclusion:

The discomfort stemmed from the truth they were expressing, from their perspective.

I recognized that honesty was valuable, as it allowed me to understand how my actions could convey messages inappropriately or even acknowledge that I was wrong on certain occasions. I also understood that sometimes it was just different points of view, and I didn't have to take it personally.

HOW DO YOU FEEL WHEN PEOPLE ARE HONEST WITH YOU?

HONESTY IS PRICELESS.

Nowadays, I am immensely grateful to all the people who had the courage to be honest with me. Although we may not always hear what we want to hear, sincerity is an invaluable treasure.

Remember, honesty creates powerful bonds between people, and if we are authentic with others, we will receive the same in return.

And this is priceless!

CHAPTER 5
PRINCIPLE 4: TEAMWORK GAME

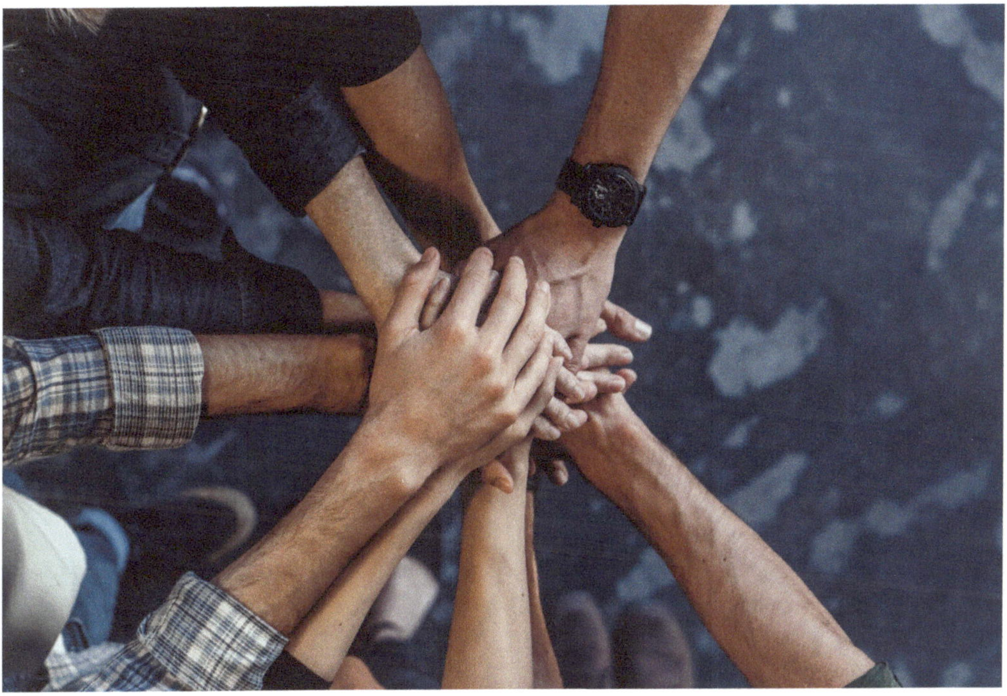

Become the best player in life by working as a team!

We are always immersed in society and, therefore, are innate players in the game of life. Now,

How can we be the best players on this grand stage?

HOW IS YOUR TEAMWORK?

The answer is simple: by working as a team. Whether in a company, family, or another community, it is crucial that we have the ability to be part of the team and play together.

Each of us is an important link in the community we live in, and as such, we must contribute to teamwork by respecting and enforcing established rules. We cannot expect to act at our convenience without considering others; it is vital to learn how to collaborate as a team.

If we seek to ensure safety in our companies and our profession, we must be part of the winning team. **But how do we know if we are on that team?**

It's simple: let's evaluate what we do for others and analyze the results. If we see that we need to improve, let's face the situation and seek progress so that in the next evaluation period, we can be the best.

The teamwork game is fundamental in any organization because it depends on maintaining a consistently high level of safety. To achieve good teamwork, it is important to eliminate any roughness that may exist among colleagues.

If you want to solve problems as a team, you can do so by following the method I implemented on numerous occasions. First, I would gather everyone at a working table with clear objectives to clarify the situation at hand.

WORK MEETINGS.

Before presenting my possible solution, I would lead the team to identify the root cause of the problem through questions. At first, there might be some tension in the room, but as the conversation progressed, the tension decreased simply because people felt heard.

Once we had identified the presented situations, we started brainstorming where everyone contributed to finding solutions. This process made everyone feel involved and committed to addressing the situation. Often, people came up with excellent ideas that significantly contributed to solving the problem.

We didn't always reach the best solution, as it depended on the specific case we were analyzing, but in approximately 80% to 90% of cases, we found a solution together and ensured it wouldn't repeat.

This action of involving everyone in the search for solutions has great power, as everyone becomes part of the solution and commits to fulfilling the agreements since they were the ones who contributed to creating them.

Remember that the idea is not to stand out individually but to make the team the winner. Whether it's the work team or the organization as a whole, the key is to join forces to achieve success.

Message to the Company

If you want to generate a genuine safety culture in your company, you must pass the ball to your employees and make them understand that they are the ones creating safety solutions. Simply giving orders to be followed to the letter is not enough; safety goes beyond that; it must permeate the BEING of each person.

True safety in the professional realm is achieved when it is internalized and incorporated into every action, in such a way that it is not necessary to have a guard ensuring that things are done correctly. Safety is measured by how people act when no one is watching. It is crucial that as an organization and higher levels, we instill confidence in our teams to ensure a safe game.

Our Goal: Safe Flights at All Times

As a company, it is crucial to ensure that each flight is conducted with high safety standards. However, this is only achieved if we trust our team to report everything that happens within the processes and areas. The key lies in how we communicate from the CEO to the middle levels and, from there, downward.

By fostering a culture of openness and trust, where every team member feels comfortable sharing relevant information for safety, we will be building a solid foundation to ensure safety in all aspects of our company.

Remember that safety is everyone's responsibility and depends on how we cultivate an environment of trust and collaboration, where each person feels actively involved in the process and is committed to the common goal: safe flights at all times.
That is the key to success!

HONESTY IS PRICELESS.

Nowadays, I am immensely grateful to all the people who had the courage to be honest with me. Although we may not always hear what we want to, honesty is an invaluable treasure.

Remember, honesty creates powerful bonds between people, and if we are authentic with others, we will receive the same in return.

And this is priceless!

WWW.AVIACIONAEROED.COM

The difference between where you are today and where you'll be five years from now is in the quality of books you read. - Jim Rohn

CHAPTER 6
PRINCIPLE 5: VISION WITH CONTINUOUS LEARNING

Develop skills to be an outstanding professional!

If we aspire to stand out as quality professionals, we must commit to continuous learning and stay at the forefront of our professions. However, it is also crucial to develop skills that allow us to improve every day in all aspects of our lives.

IDENTIFY WHAT YOU LEARN EVERY DAY.

Throughout my journey, I have learned and applied various skills that have transformed my perspective and significantly enriched my life:

Money Management: Often, we fall into the trap of the rat race without valuing the importance of learning to manage money intelligently. We all need to acquire financial knowledge, even as employees, to be able to grow economically. An interesting fact I discovered is that wealthy people enjoy what they do and don't worry about the paycheck because they have created assets that generate passive income. Although I am still in constant learning, this knowledge motivated me to take action and reorganize my financial life. Today, I feel happy as I am building my path to financial freedom.

Body Care: I understood that to be the best in my profession, I had to offer the best to my body. So, I incorporated healthy habits into my life, prioritized proper nutrition, and regular exercise. This has provided me with the strength and endurance necessary to face any challenge in my profession.

Adoption of Healthy Habits: Adopting healthy habits may not be easy, but it is not impossible either. It depends on the focus we give to our lives. If we want to be top-notch professionals, we must abandon mediocre attitudes and take action. Taking care of our health, investing in ourselves, and paying attention to our nutrition are essential steps to live a fulfilling life and avoid problems in the future.

IS YOUR BODY READY TO FACE ANY SITUATION?

HAVING THE INFORMATION WITHOUT TAKING ACTION IS OF NO USE.

It is crucial to remember that action is the key to break free from the normalcy imposed by society. **By immersing ourselves in a constant learning zone and stepping out of our comfort zone,** we cease to be ordinary professionals. I always repeat these words to myself when I feel I am deviating from the path to the excellence I want to achieve.

These are just a few examples of the skills I have incorporated into my life, and I'm sure I can share more with you if you wish. The important thing is that we all have the opportunity to transform into exceptional professionals and reach our full potential.

It's time to take action and become the best!

MESSAGE TO THE COMPANY

Upper management must have a training plan for positions, ensuring that staff is always well-trained and up-to-date. This way, operational safety is guaranteed.

If you want to foster a culture of safety in the company, it is crucial to consistently train the personnel. However, it's important to note that the training should be specific and personalized, tailored to the needs of each group. Training pilots and co-pilots is not the same as training flight attendants or technicians. The way we convey the message to all staff will be key to building a strong culture.

Upper management should have a training plan for positions, ensuring that the staff is always trained and updated. This way, safety in operations is guaranteed.

If you want to promote a culture of safety in the company, it is crucial to consistently train the personnel. However, it is important to emphasize that the training should be specific and personalized, tailored to the needs of each group. Training pilots and co-pilots is not the same as training flight attendants or technicians. The way we convey the message to all staff will be key to building a strong culture.

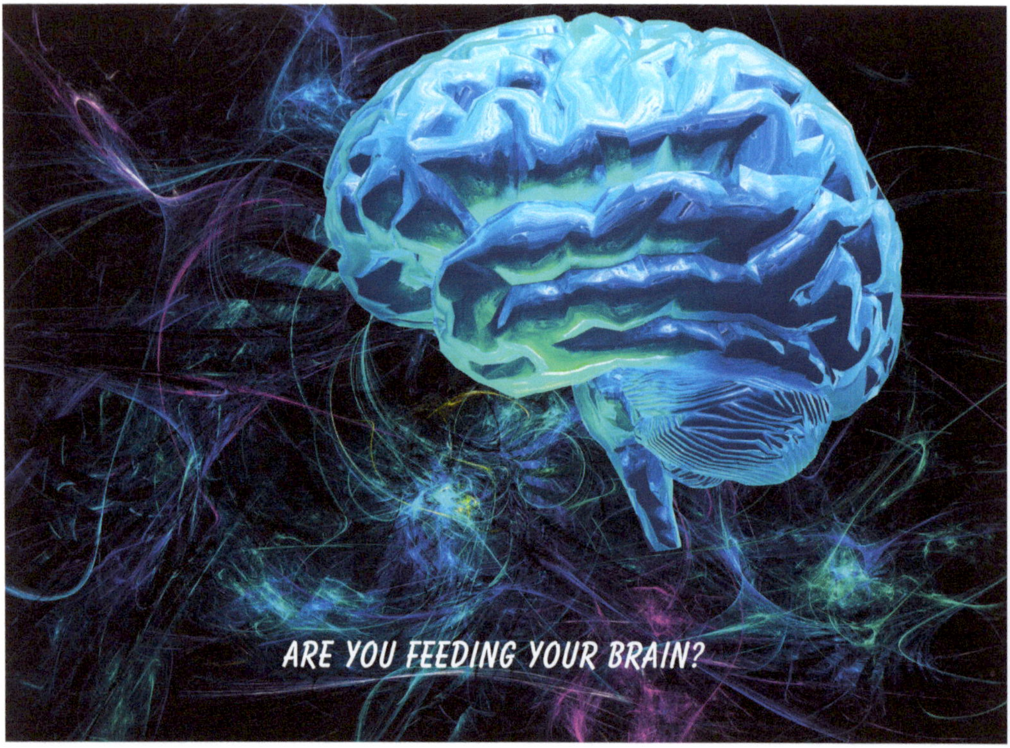

ARE YOU FEEDING YOUR BRAIN?

Our mind is one of our most valuable assets, and we must consciously take care of it. Everything we read, hear, and do contributes to who we are and how we navigate through life.

It's important to note that my experience has allowed me to develop specialized training to raise awareness of safety in individuals.

WWW.AVIACIONAEROED.COM

I believe everyone should study ants. They have an amazing four-part philosophy. Never give up, look ahead, stay positive, and do all you can. - Jim Rohn

CHAPTER 7
PRINCIPLE 6: INDIVIDUAL RESPONSIBILITY

When I started taking responsibility for my life and surroundings, I understood that my success depended entirely on me. I realized I had to take charge of the necessary adjustments. By taking responsibility for the information in my mind, I became a differentiating professional. After all, no one would read for me, nor exercise for my body. The reality is that no one else would take care of my affairs, just as it will happen with you; no one will do things for you.

DON'T WAIT FOR SOMEONE TO DO SOMETHING FOR YOU.

We are solely responsible for our lives and how we live them. In my personal and professional life, I decided to take action and face the consequences of my actions, which built people's confidence in my work. They knew that if I made mistakes, I would acknowledge and correct them without excuses.

This principle of responsibility is closely linked to other aspects of our lives, as our attitude determines the level of responsibility we will uphold.

To facilitate my professional development, I committed to following each of the principles I mentioned, enabling me to handle multiple responsibilities successfully as I was adequately prepared.

Remember the wise words of John C. Maxwell:

The greatest day in your life and mine is when we take total responsibility for our attitudes. That's the day we truly grow.

To be successful in both personal and professional aspects, we must approach each commitment as our own, exercising responsibility at all times.

When I understood this, my life began to change dramatically. I became wiser by taking control of my own future.

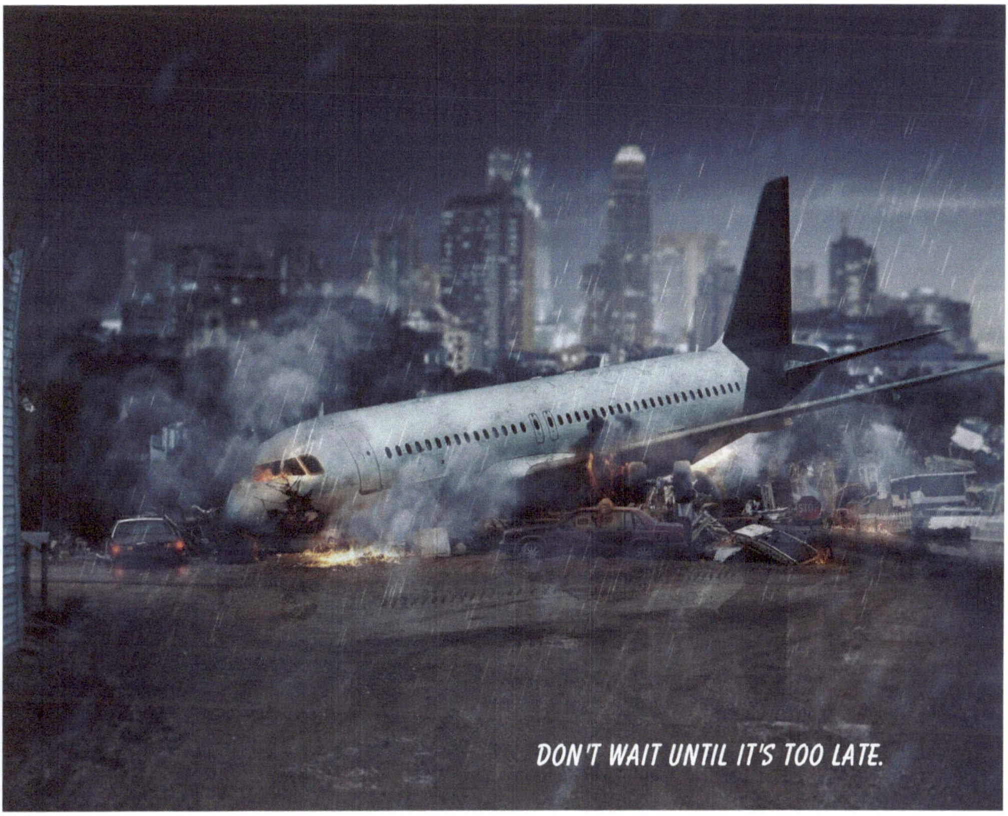

DON'T WAIT UNTIL IT'S TOO LATE.

Analyzing research on aviation events, accidents, and incidents, I noticed that there are often failures in processes that have been present for a long time. If those responsible for these errors had exposed and addressed them in a timely manner, losses of money, time, resources, aircraft, and even lives could have been avoided.

Unfortunately, in many instances, it is only after experiencing such losses that someone finally takes responsibility.

So, I want to invite everyone reading this to take this message into their lives and workplaces: let's assume our responsibility at all times, let's not wait until we suffer losses to correct things, as it may be too late.

Responsibility is the key to staying on the path of professional and personal success, preserving our good name and reputation. It's time to take control of our lives and destiny, and thereby grow and prosper significantly.

TAKE THE NECESSARY ACTION FOR YOUR LIFE.

MESSAGE TO THE COMPANY

Within organizations, people often fear telling the truth and taking responsibility for possible consequences. However, it is at this point where upper management must provide support to their employees and teach them that in aviation, as in any field, it is crucial to be accountable for our actions.

Treating people as responsible professionals is the only way to foster a true culture of responsibility. We must give them the opportunity to speak up and explain what is happening, as well as allow them to correct potential mistakes in their work.

When seeking safety in companies, it is essential to build trust among employees so that they willingly share everything related to the operation. While some may argue that this should be an obligation of the employees, it is important to question whether they are truly sharing everything solely out of obligation.

Creating a conducive environment for communication and trust is necessary, where developments are addressed effectively, allowing workers to proactively assume their responsibility.

Upper management should set an example of responsibility in all aspects of the operation, understanding that compliance with aviation-related norms, laws, and regulations is indispensable.

TODO EL PERSONAL DEBE CONFIAR EN LA EMPRESA

SET AN EXAMPLE FOR YOUR WORKERS

Fulfilling responsibilities as responsible companies is the key for employees to feel motivated to take responsibility for their actions.

Let's remember that the example we set will influence others. If we demonstrate a genuine commitment to responsibility, we will positively impact the attitude of the entire team, thereby creating a strong and sustainable culture of responsibility.

WWW.AVIACIONAEROED.COM

Words do two major things: They provide food for the mind and create light for understanding and awareness. - Jim Rohn

CHAPTER 8
PRINCIPLE 7: EFFECTIVE COMMUNICATION

The Power of **Effective Communication in Aviation**

Communication is an age-old powerful principle that has been key throughout the history of humanity. While we all believe we master it perfectly, I admit that initially, I struggled to understand its true significance.

Communication has always been essential in the workplace, especially in the aviation industry, where it is crucial to ensure safe air operations. However, surprisingly, as aviation professionals, we sometimes fall short in this crucial aspect.

The lack of effective communication has been identified as one of the leading contributing factors in aviation accidents, as it creates deficiencies in processes and leads to negative outcomes in aviation.

We depend on communication to ensure that everyone understands what we should do and what we expect others to do. Communicating effectively is an ongoing task that requires dedication, and we cannot assume that simply by talking, we will be understood or understand others.

Often, we fall into the trap of believing that others share our perspective simply because it's clear in our own minds.

It's interesting how many unresolved situations could be resolved with a simple conversation, but we resist doing so because we assume that others should already know what we are thinking.

Now, understanding the relevance of communication in aviation operations, I realized that I needed to learn how to address difficult conversations, especially when ego plays a significant role.

This principle taught me that if I wanted to be an outstanding professional, I first had to learn to **LISTEN.**

LISTEN TO UNDERSTAND THE OTHER PERSON'S PERSPECTIVE.

We cannot know the situations that the other person is going through or has experienced to be at the point where they are. If we seek to distinguish ourselves as professionals, we must be capable of understanding others and, if necessary, ask questions until the message is completely clear. It is a duty we must fulfill.

Communication is a fundamental pillar in our professions, affecting all the activities we carry out daily. Let's never assume that someone understands a situation or should be aware. We lose nothing by asking or clarifying.

Communication is the backbone that ensures that aviation operations always remain in a safe zone. It's time to harness its power and use it to make a difference.

MESSAGE TO THE COMPANY

At the highest levels of organizations, there is a significant responsibility regarding communication. It involves teaching the art of listening, not merely restricting oneself to issuing orders constantly.

It is crucial to remember that safety is a shared responsibility, especially in aviation. To fulfill this mission, it is essential to learn to listen to what is happening within processes and to convey the message appropriately.

The way we convey the message will determine how employees carry out their tasks. In my experience, I have witnessed various organizations where employees feel dissatisfied due to the way they are treated. Sometimes, this dissatisfaction is not solely attributable to senior management but also to certain middle levels that fail to recognize how our communication can positively or negatively impact the actions of workers.

Faced with this flaw in organizational processes, I have focused on strengthening the way messages are conveyed and maintaining fluid communication in all directions within the organization. It is from this point that the foundation of a robust safety culture begins to take shape. Effective communication is the cornerstone of this process.

CONNECT WITH THE STAFF

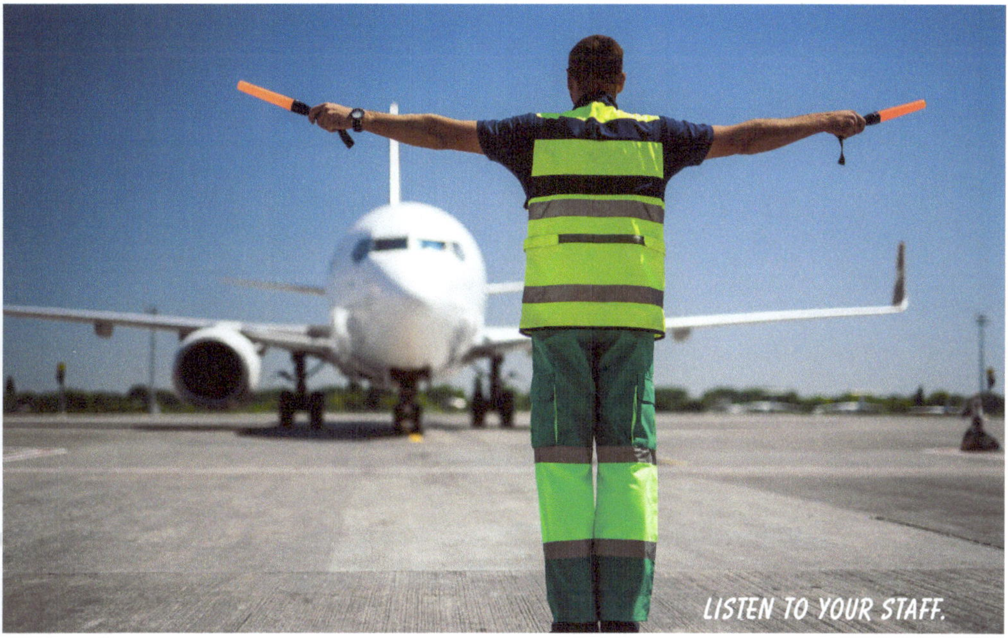

By creating a safety culture based on open and responsive communication, we foster trust and commitment among all team members. This culture directly impacts the efficiency of operations and the overall well-being of the organization.

It is time to recognize the importance of communication as a key factor for success in aviation and any other industry. By learning to listen and convey messages effectively, we can achieve greater excellence in our operations and create a more positive and productive work environment for everyone. Communication is the foundation for building a safer and more successful future.

CHAPTER 9
PILLAR OF ACTION

Do you want to experience real growth, Edwin?

This was a question I asked myself when I began incorporating these principles into my life. I realized that no principle would work as I wanted unless I added a fundamental ingredient to everything I did. Yes, I'm talking about

ATTITUDE.

While it is true that with the principles I shared in this book, I was able to organize my personal and professional life, there was something that made the difference in my transformation: attitude. Learning to face each situation and apply each principle with the right attitude was crucial.

The attitude with which we receive and perceive every moment of our lives is the key to making everything simpler. Since I adopted a positive and winning attitude, everything in my life began to change, and I am confident the same will happen for you.

Allow me to share a revelation with you. When I started applying the principles to improve as a person and professional, I discovered aspects of myself that I was unaware of. I thought I was acting appropriately, but as I delved into the art of facing different work scenarios, I realized how my attitude was influencing my life.

Exercise played a significant role in this process. When I started practicing high-intensity exercises before facing my day, I noticed a change within myself. I'm not an expert in energy-related matters, but I understood that my attitude began to shift. Pending situations became more manageable, and I became less reactive in my daily activities.

All the information I share in this book and on my social media platforms aims to positively impact you. I want you to know about my life experience, and if it's helpful to you, incorporate it into your own life. I don't intend to change your life as I did with mine; that responsibility is yours. I only recommend that you live each day working to be the best version of yourself. This isn't just about being the best professional but about being the best 'you' compared to your past 'self.' That's the true challenge we all face.

If you are still reading these lines, I encourage you to approach all the information with the best possible attitude and to incorporate these principles to generate great value in your life.

Life is too short to waste on things that do not passion us. Let's support those who wish to learn from us and share our knowledge with those who need it. Each of us has a valuable story that could inspire someone in the world.

Now, let's talk about how attitude affects safety in aviation operations. Our way of living, our experiences, and our values influence the safety of aviation operations. The attitude with which each aviation professional faces situations is crucial to ensuring a safe flight.

Imagine this situation: the plane is ready to take off, but the captain announces that due to technical issues, there will be a delay. Many passengers are upset after waiting for over two hours in the boarding area. However, the captain steps out of the cockpit and stands in the aisle with firmness and empathy. With a compassionate voice and a request for everyone's cooperation, he communicates the situation and asks for understanding.

The captain's message, conveyed with a positive attitude, builds trust among the passengers. Despite the delay, everyone accepts the situation and organizes their time calmly and collaboratively.

This story demonstrates how a positive attitude can make a difference in any situation, even in an inconvenience. When professionals convey positive energy, we change the atmosphere and can face any challenge with a different perspective.

OUR WAY OF ACTING DETERMINES THE ACTIONS OF OTHER PEOPLE.

AND TODAY, HOW IS YOUR ATTITUDE?

Like a muscle, attitude is trained and strengthened day by day. Living with a positive attitude is a matter of decision. By adjusting our environment and focus, we can convey more than 80% positive energy in our interactions.

Understanding that we are responsible for our attitude and can positively influence others empowers us. In any circumstance, we can choose to give our best and not allow negativity to affect us.

In summary, attitude is essential in aviation and in life. With a **positive attitude**, we transform operational safety and create a more harmonious environment. Let's inspire others to live with a positive attitude and contribute to creating an environment where everyone can fly high and safely.

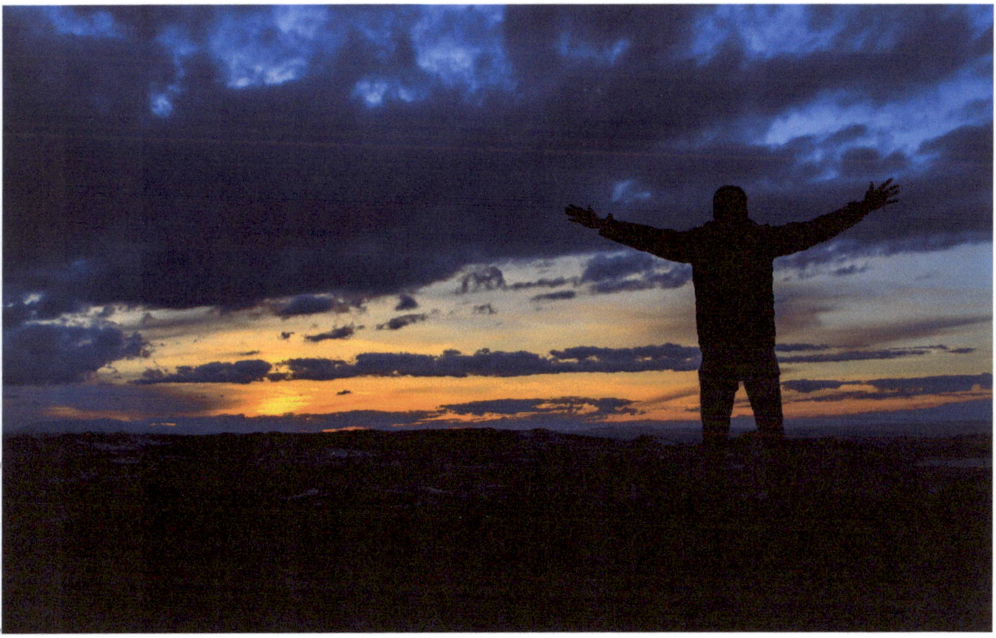

MESSAGE FOR YOU

ATTITUDE is one of the fundamental pillars supporting the principles presented in this book.

Remember, we only have one life, and every minute counts...

What do we gain by staying upset all the time?

Each of us has the power to improve our life.

ATTITUDE is the key to keeping safety high and is the foundation that holds up the principles described in this book.

CHAPTER 10
PILLAR OF VIRTUE

The only difference between an ordinary professional and an extraordinary professional is how they perform everyday tasks exceptionally. All of this is achieved through a virtue that few develop but is essential if we want to stand out in our field.

I'm talking about actions that very few people are willing to take. I'm talking about actions that can completely change our lives, and I am certain of it.

If you incorporate this virtue into your life, everything around you will automatically change. This virtue, present in all successful individuals, is

DISCIPLINE.

Discipline is crucial in our professional lives, much like how the wind is vital for an airplane to fly and reach its destination. It involves performing a series of daily actions to keep everything around us in the shape we desire.

Ordinary professionals are such because they continue to behave like common people. They don't read books of knowledge, don't care about their health, and expect to receive everything without giving anything in return.

The EGO is also an obstacle that keeps some people in an undesirable place. In the aviation industry, this is seen in those who believe that their title or position exempts them from learning more.

In my case, I admit that I didn't incorporate proper disciplines into my life before, and that prevented me from progressing rapidly. Today, I understand the importance of this virtue and want to share it with you.

I have replaced negative habits with more constructive ones. Now, I watch documentaries and movies with a focus on personal growth instead of Netflix series that add no value. I have also minimized my alcohol consumption and use my time on social media to convey useful messages.

START MAKING SMALL ADJUSTMENTS TO YOUR LIFE, EVERYTHING WILL CHANGE.

I stopped making excuses and started reading more, understanding that time and mind are our greatest assets. Persistence and discipline are crucial to achieving our goals both professionally and personally.

Setting goals is essential. I created a list of 100 goals I wanted to achieve before I die, and this material you now have in your hands was one of them. Organize your life plan, outline your goals, and work diligently towards them.

Remember that discipline is doing things every day, even if you don't feel like it, knowing it's necessary to achieve bigger goals. If you aspire to be an extraordinary professional, discipline is the key.

CHAPTER 11
THE LAST

Finally, let's talk about the importance of incorporating safety into our professions.

It is crucial to understand that every action we take has an impact on the safety of those around us. Safety is not just a task to fulfill but a working tool for the entire organization.

It is unfortunate that safety is sometimes relegated for economic interests. Nevertheless, it is the responsibility of everyone, regardless of their position, to prioritize safety in our actions.

Introducing safety into organizations is not an easy task, but it is essential. The question we all must ask ourselves is whether it is more cost-effective to invest in safety and prevent accidents or face the consequences of an accident.

When we sign our name in our aviation profession, we assume the responsibility for safety on every flight. Aviation is our passion, and we must always maintain a balance between production and protection.

The phrase **LIFE IN A SINGLE FLIGHT** encapsulates the importance of conducting our actions with responsibility and safety. Let's avoid situations that could lead to negative consequences and instead work towards maintaining a safe and protected environment.

Money and luxuries are meaningless if we cannot enjoy them fully. True wealth lies in living with integrity and responsibility. We can always contribute to others by sharing our knowledge and experience.

We always attract what we become, so we must take actions to be the best version of ourselves.

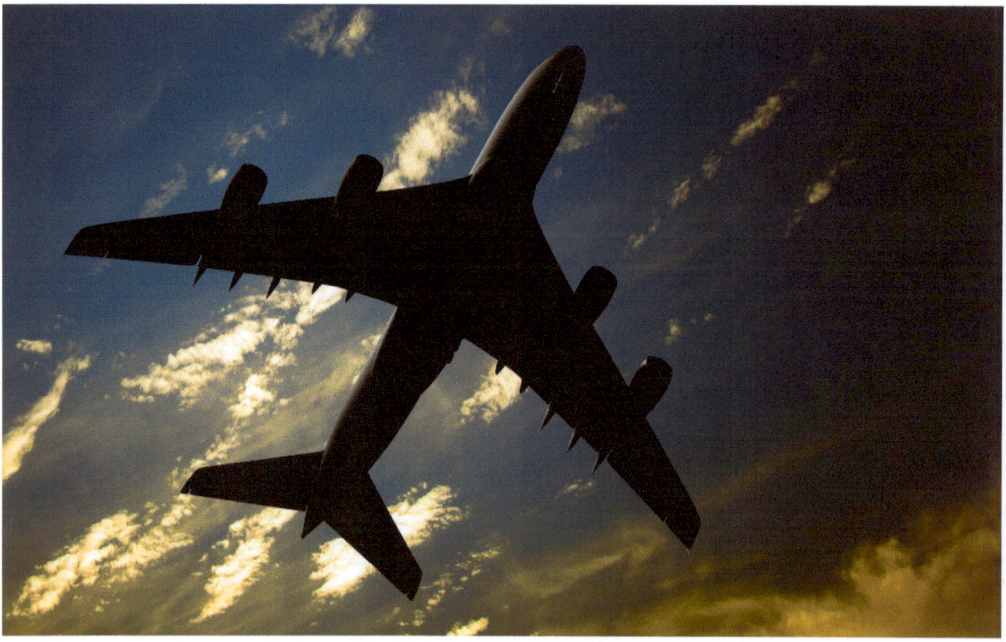

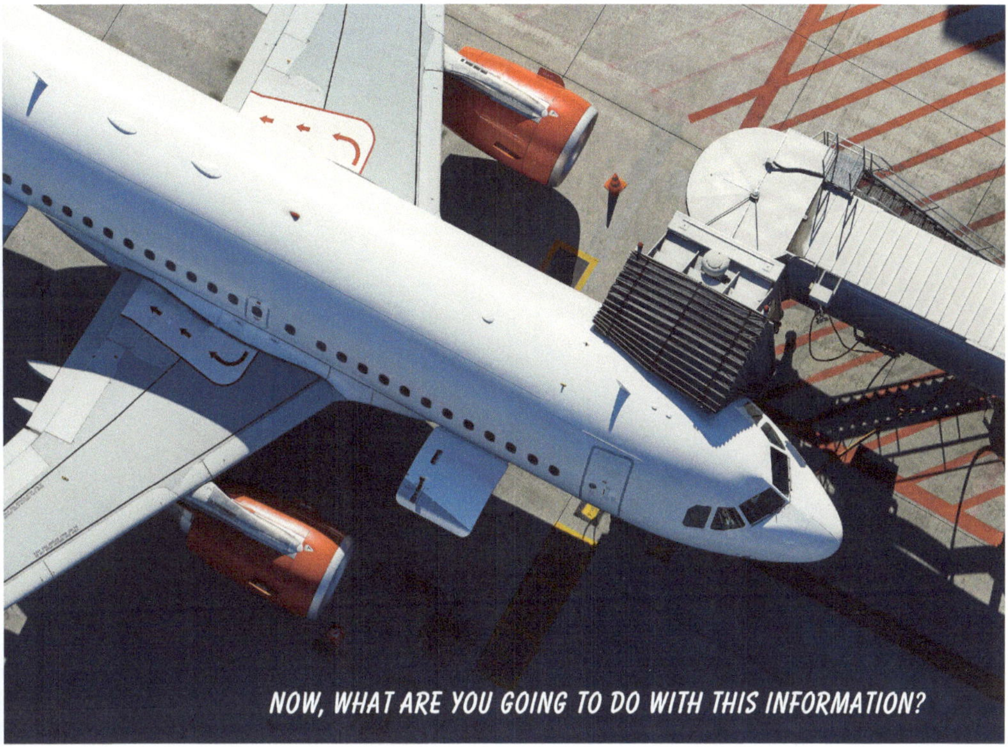

NOW, WHAT ARE YOU GOING TO DO WITH THIS INFORMATION?

Focus on being extraordinary in what you do, stand out, and be impossible to ignore. You're not here to be just another; you're here to win.

Now is the time to implement what you have learned. Information without action has no value. The responsibility is in your hands.

Thank you for reading this book and sharing a part of my life with you.

www.ingramcontent.com/pod-product-compliance
Lightning Source LLC
Chambersburg PA
CBHW040317220526
45473CB00009B/2473